DINOSAUR SHAMAN

BY
MARK
SCHULTZ

NINE TALES FROM THE XENOZOIC AGE

Princeton KITCHEN SINK PRESS Wisconsin

Thanks to my colleagues at Kitchen Sink Press for their continued fine efforts in support of Xenozoic Tales.

Thanks to that legendary Texan Jack Jackson for his thoughtful introduction.

And a special thanks to Steve Stiles whose skillful interpretations of my scripts add so much richness and life to the Xenozoic Age.

— Mark Schultz

Mark Schultz: creator, writer, artist, *Xenozoic Tales*; **Steve Stiles**: artist, additional stories; **Denise Prowell**: letterer and front cover color; **Ray Fehrenbach**: art director and back cover color; **Jan Manweiler** and **Christi Scholl**: production; **Jack Jackson**: introduction; **Dave Schreiner**: editor; **Denis Kitchen**: publisher.

ISBN 0-87816-117-1 (hardcover)
ISBN 0-87816-118-X (softcover)

Library of Congress Cataloging-in-Publication Data

Schultz, Mark. 1955 -
Dinosaur shaman : nine tales from the Xenozoic Age / by Mark Schultz
p. cm.
First published in Xenozoic tales, issues 5-8.
ISBN 0-87816-117-1 : $29.95 -- ISBN 0-87816-118-X : $14.95
I. Title
PN6728.D56S38 1990
741.5'973--dc20 90-21624
CIP

Table of Contents

Introduction

by Jack Jackson

Every once in a while, someone like Mark Schultz bursts upon the comics scene. Because it's relatively rare for a new artist to emerge already possessed of masterful technique, such an occasion is a joyful one to those of us who love this medium. This sense of wonderment is compounded when the artist also writes his own stuff, as Mark does. It means that we can look forward to years of top-notch entertainment as his talent—both as artist and storyteller—grows and matures.

This book is the second collection of Mark's work, which appears in the comic book *Xenozoic Tales*. Those of us who have followed the series or read the first collected volume, *Cadillacs & Dinosaurs*, will notice how rapidly his style is developing. Mark's pilot strip from the Xenozoic saga, "Mammoth Pitfall," done in 1986, is surprisingly good for a first effort. All the essentials are there, even if the art is a bit crude and overtly derivative—compared to his current work, that is. We all have to start somewhere, and Mark Schultz started by studying the masters, comics greats like Frank Frazetta, Al Williamson, Wally Wood, Harvey Kurtzman and Will Eisner. Now, after only a couple of years, he already ranks with the best of them. To be able to come so far so fast is truly amazing, and if this progress reflects the depth of Mark's genius, we can anticipate some great stuff from him in the years ahead. Steve Stiles, who works from Mark's scripts in the backup stories in *Xenozoic Tales*, is also represented in this collection, as he was in *Cadillacs & Dinosaurs*. Steve continues his sympathetic treatment of Mark's vision in these stories, and is a real asset to the *Xenozoic* series.

Like most cartoonists I know, Mark started drawing (dinosaurs, naturally) at an early age. Unlike most of them, he went on to study art in college and actually graduated with a degree in fine art. Predictably, art school did little about furthering Mark's interest in comics, trying instead to channel his talent into painting for the manipulative gallery scene. During his college years (1973-1977), he was aware of some of the undergrounds but not tempted to try his hand. Mainly, the comix movement demonstrated to him that it was still possible to write your own material—whatever you desired—and that you did not have to "buy" the piecemeal, assembly line approach pushed by the giant companies. This came as something of a revelation, one that Mark Schultz did not forget.

In the early '80s, after some unsatisfying illustration and advertising work, Mark reached a turning point in his life. EC fans will love this story: He took several boxes of his old Marvel and DC comics down to the local comics shop and swapped them for an assortment of about 50 ECs, which the shopowner by chance had just obtained. After carefully studying the books, admiring their writing and art, allowing the EC magic to work its spell, Mark realized that he'd found his calling—to make books as spellbinding and great as these!

And so he has. Building upon the EC science fiction-fantasy legacy, Mark has taken one of its most potent themes—that of a civilization struggling to rise from its own ashes—and given it a twist that we cannot ignore as we witness man's constant meddling with the "natural order" of things and the rapid erosion of our ecosystem as a result.

Ever present in his stories is the unifying thread of nature's delicate balance, in which all life forms are interconnected. Over this looms the specter of a technology that got out of hand, almost destroying and profoundly altering the planet. Science's dark secrets, when unlocked, still pose a threat to the survivors. Now comes the central question: How should the survivors of a near-apocalypse regard such knowledge? Should they embrace it and attempt to learn from past mistakes, or should they shun it and fight to keep it inaccessible from those who—either through malice or stupidity—might finish the job?

Jack Jackson created one of the first underground comix, God Nose, *and he is one of the premier artists that came from the underground comix movement. His more recent historical graphic novels include* Comanche Moon *and* Los Tejanos. *He recently published a historical fantasy called* Secret of San Saba.

This quandary sets the stage for Mark's characters and adds dimension to their realization, the "fleshing out" process by which we come to recognize them and respond to them as special people. A word of caution: Do not be lulled into thinking that his Jack Tenrec is as simple a character as Indiana Jones, for "Cadillac Jack" grapples with existential issues wholly different from those that motivate Jones and the old adventure genre from which he sprang. Regardless of what is taking place in any particular strip, however mundane, Mark always manages to keep the cosmic proportions of his tale in view. In this respect his work deals with themes similar to those addressed by the ECs and the undergrounds in their finer moments.

A notch above whiz-bang entertainment, Schultz's futuristic world is "real," down to the smallest detail. Out-of-whack, fantastic, and yet real—not an easy trick to pull off with mere lines on paper. In truth, it could only be done with *our* involvement and this is where Mark shows his mastery, because his art beguiles us into entering this strange world he's created. Not since the EC work of Frazetta, Williamson and Wood have we seen such compelling, exquisite panels and felt the urge to "let go" and be swept away by an artist's vision. After the EC era, those who could transport us in such a manner have been few and far between (Jack Kirby, Moebius, and Richard Corben come to mind; for me, anyway).

But nobody puts it all together quite like Mark Schultz, or has produced a sustained effort like *Xenozoic Tales*. Reading his stuff is akin to a stroll through the Golden Age of adventure comics, and it does my heart good to know that the art form is not only alive and well but has evolved into something more powerful philosophically. Further, after the X-rated excesses of the undergrounds (for which I accept my part of the credit), it's nice to see someone who understands sexual tension and knows how to exploit it without even needing to undress his characters. That is a miracle in itself!

What we have here is a rip-snorting adventure comic drawn with ever-increasing beauty; a serious commentary on our fitness for survival as a species, but set in an evolutionary fantasyland; an incredibly sexy strip, yet suitable for kids; and—I predict—a saga that will some day be ranked among the top comic book creations of the 20th century. So enjoy it, folks; there's not many others like it in this day and age, where everything from diapers to art is designed to be thrown away. Mark has made sure we'll want to keep this one.

Pages from the Mark Schultz sketchbook...

Presenting 10 pages of pencil and ink exercises, roughs, thumbnails, convention sketches and other material produced by Mark Schultz, creator of *Xenozoic Tales*. Schultz recently won a Harvey Award as Best Artist of 1989. The Harvey Awards are nominated and voted upon by comic book professionals, who seem to agree that Schultz is one of the 'hottest' new talents to arrive on the scene in years. Schultz works hard to improve his art, rendering detailed pencils and then striving to carry through that detail with his inking. The result is one of the most beautiful graphic works to be produced in years, *Xenozoic Tales*...

BELOW: An early (1983) attempt at the concept that evolved into XENOZOIC TALES. The characters are already in place, although their costumes will change dramatically.

Early cover rough for **DINOSAUR SHAMAN.**

On the next three pages, various sketches for a proposed Hannah Dundee porcelain statue yet to be sculpted, and other drawings featuring the XENOZOIC TALES heroine.

Working out figure poses.

Sketches for proposed serigraph.

On the previous two pages, penciled art for pages 3 and 4 of "In the Dreamtime," from XT 8, and the last story reprinted in this book. On this page, thumbnail sketches of the same two pages, and below, a rejected design for a t-shirt.

EXCURSION
WELCOME BACK TO THE XENOZOIC!
IT IS THE SECOND DAY FOLLOWING THEIR ADVENTURE IN THE CITY'S TREASURE VAULTS. NOW JACK TENREC MOVES TOWARD AN EARLY MORNING CONFRONTATION WITH HANNAH DUNDEE. SO DOES THE WEB OF CIRCUMSTANCE CONTINUALLY DRAW THE UNSUSPECTING TOGETHER...
BAM
BAM
BAM
MARK SCHULTZ © 1988
1

C'MON... I KNOW YOU'RE IN THERE...
BAM
BAM
JACK?
FOR CRYIN' OUT LOUD! I JUST GOT TO BED...
UP ALL NIGHT, HUH? AS IT APPEARS YOU'RE ALONE, I'LL BET YOU'VE FOUND A GOOD BOOK!
WHAT'S THAT SUPPOSED TO MEAN?
IT MEANS THAT A TRUSTWORTHY SOURCE HAS INFORMED ME THAT A BOOK WAS STOLEN FROM THE LIBRARY VAULT DURING OUR RECENT JUNKET.
COINCIDENTALLY, A LATE 20th CENTURY SCIENCE BOOK.
SO WHY HASN'T THE GOVERNING COUNCIL COME TO ME WITH THIS CHARGE?
MAYBE THEY'RE TOO EMBARRASSED TO DEAL WITH YOU THEMSELVES.
YOU MEAN WITH ALL THOSE VOLUMES LOST IN THE FLOOD, THEY THINK I STOLE ONE LOUSY BOOK?
WHAT'S THIS?
IT'S A TRITON NET I'M WORKING ON. PART OF THE TRADE AGREEMENT WE WORKED OUT CALLS FOR THE WASSOON TO HELP IMPROVE YOUR TRIBE'S ABILITY TO GATHER FOOD FROM THE SEA. THERE ARE RESOURCES DOWN THERE YOU'RE IGNORING.
2

BUT... TRITONS AREN'T EDIBLE!
SHOWS HOW MUCH YOU KNOW.
SO ENLIGHTEN ME.
LET'S TAKE A BOAT OUT. YOU CAN DEMONSTRATE YOUR TRITON NETTING TECHNIQUES TO ME AND WE'LL HAVE A CHANCE TO ...er... TALK.
YOU OWE ME FOR THE LIBRARY TRIP...
WELL, I HADN'T PLANNED ON SPENDING THE DAY BEING INTERROGATED... BUT IF IT'LL GET YOU OUT OF MY QUARTERS...
GIVE ME A MINUTE TO GET READY.
BELIEVE ME, YOU'RE NOT GOING TO FIND IT HERE...
...CREEP!
SOON, AT THE DOCKS...
...JUST A FEW MORE OF THESE AND WE'LL BE READY...
AMBASSADOR DUNDEE! TENREC!
3

WE THOUGHT YOU'D LIKE TO KNOW THAT THE...uh... MOLES HAVE ALL BEEN ROTATED TO THE SURFACE. THEY'RE NOT VERY HAPPY TO BE TOPSIDE AGAIN, BUT THEY'LL HAVE TO LEARN TO ADJUST. SEEMS THEY BECAME A LITTLE... INTROVERTED DOWN THERE.
BY THE WAY, TENREC...ONE OF THE MOLES... USHIJIMA... MENTIONED HE GAVE YOU SOME INFORMATION. YOU TOLD HIM YOU'D BRING IT TO THE COUNCIL'S ATTENTION?
!
IT'S...IT'S NOTHING THAT CAN'T WAIT, GOVERNOR DAHLGREN.
YOU TWO HAVE A SAFE TRIP...
WHEREVER YOU'RE HEADED.
4

SO YOU WERE...ummm... TALKING WITH USHIJIMA?
DON'T PLAY CUTE. YOU KNOW I WAS.
WHO ELSE COULD HAVE SEEN YOU SWIPE THAT BOOK?
AND YOU BELIEVE HIM!
WHAT IS THIS... BLACKMAIL? YOU THINK I'LL COOL MY OPPOSITION TO YOUR ANTI-POACHING JIHAD?
BAH! THAT'S NO PROBLEM!
NO, YOUR SECRET'S SAFE WITH ME!
WHAT SECRET? I'VE ADMITTED NOTHING!
...THIS IN SPITE OF THE FACT THAT I'M UP TO MY ARMPITS IN SLITHER GUANO WITH THE GOVERNORS...
WHICH IS WHY WE'RE BEING TAILED.
THE COUNCIL'S NEVER CARED MUCH FOR ME, BUT NOW BETWEEN GORGOSTAMOS' "MYSTERIOUS" DEMISE AND OUR LITTLE TOUR OF THE VAULTS, I'M AFRAID THEY REGARD ALL OUR ACTIONS AS HIGHLY SUSPICIOUS.
5

YOU'D THINK THEY'D BE HAPPY WE SAVED THEIR LIBRARY.
THEIR CONCERN IS THAT AN UNAUTHORIZED PERSON GAINED ACCESS...
YOU CITY DWELLERS... YOU'RE ALL FASCISTS...
WHAT THE HELL DOES "FASCISTS" MEAN?
YOU OF ALL PEOPLE SHOULD KNOW.
I DON'T KNOW WHAT YOUR COUNCIL'S SO AFRAID OF... THE WEALTH OF INFORMATION DOWN THERE COULD BE SUCH A BENEFIT TO US ALL IF...
HERE, LET ME TAKE THE TILLER FOR A WHILE. I WANT TO TELL YOU A STORY...
WE WASSOON LIVE IN A GREAT SALT FEN THAT HAS ABSORBED THE RUINS OF A CITY OF WHITE MARBLE...
"A city that, unlike your own, was completely leveled by the cataclysm..."
6

"I'm sure you already know that we are a tribe of mighty hunters and fishermen..."
"But we are also a tribe of scholars... descended directly from the great lords of the ancient world."
"As such, we preserve and pass on as much of the ancient disciplines as are remembered. Our city... unlike yours... was left with no pre-cataclysmic writings. We have composed all our texts from memory."
"Through contact with poachers driven from the north, we learned of your city's vaults...and the library.
To a people with only a spoken tradition, the knowledge in those vaults... fixed... immutable... pulls like a magnet."
"The rest you can figure out..."
I WAS CHOSEN TO OPEN RELATIONS BETWEEN OUR TRIBES...TO FORGE A TRADE AGREEMENT... BUT MOST IMPORTANT, TO NEGOTIATE FOR WASSOON ACCESS TO THE LIBRARY...
7

SO YOU KNEW OF THE LIBRARY LONG BEFORE I BROUGHT IT UP?
OF COURSE. AFTER THE COUNCIL FLAT OUT REFUSED TO EVEN LISTEN TO MY REQUEST, HOW COULD I TURN MY BACK ON YOUR OFFER? IT WAS MY ONLY CHANCE TO LEARN ANYTHING AT ALL...
YOUR COUNCIL HAS ALL THE FORESIGHT OF A SENILE SAMBUK!
IF WE WASSOON HAD ACCESS TO YOUR LIBRARY, THIS WHOLE WORLD WOULD BE MUCH IMPROVED!
IF THE WASSOON ARE ALL AS JUDGMENTAL AND AS LIGHT-FINGERED AS YOU, THEN THE WHOLE WORLD'S IN TROUBLE.
IT'S TIME YOU HAD THE WIND TAKEN OUT OF YOUR SAILS.
HANNAH RELEASES THE SHEET, THE SAIL FLUTTERS FREE, AND...
8

NOW WHY'D YOU DO THAT?
I DON'T LIKE BEING CALLED JUDGMENTAL...
YOU DON'T ARGUE WITH LIGHT-FINGERED...?
9

DRUM
DRUM
DRUM
YOU CAN STOP FOOLING AROUND NOW, JACK.
JACK?
SPLOOOSH
10

AS LONG AS I WAS WET, I THOUGHT I'D GET US SOME LUNCH... AT GREAT PERSONAL RISK I MIGHT ADD.
IT WOULD BE A GREAT PERSONAL RISK TO TRY TO EAT THESE TOUGH OLD SOOKS.
BUT THEY'LL MAKE ACCEPTABLE BAIT.
LET'S PULL INTO THE MARSH HERE, JACK... IT'S TIME TO MAKE A FISHERMAN OUT OF YOU.
THIS ALMOST LOOKS LIKE HOME!
TRITONS SPEND THEIR EARLY YEARS IN THE MARSH AND THEN COME BACK TO MOLT. THE ONES WE'LL CATCH HERE WILL BE NICE AND TENDER... GOOD EATING!
DROP ANCHOR. WE CAN TAKE THE SKIFF IN FROM HERE.
11

HERE COMES OUR CHAPERONE.
IF THEY SO MUCH AS TOUCH MY BOAT WHILE WE'RE AWAY, I'LL KNOCK THEIR HEADS IN.
WE'LL SPREAD OUR FIRST LINE OF NETS HERE.
...AH... THIS SMELLS LIKE HOME...
!
IT STINKS!
BUT SPEAKING OF YOUR HOME... WON'T YOU BE REPORTING BACK SOON?
I PROBABLY SHOULD BE... BUT NOT BEFORE I CAN LEARN A LITTLE BIT MORE ABOUT YOUR FRIENDS, THE GRITH!
I KNEW IT! I KNEW YOU'D HAVE TO BRING THEM UP!
SO WHO'S GONNA HEAR US TALKING OUT HERE?
NO ONE BUT THEM, MAYBE.
DO YOURSELF A FAVOR AND DON'T EVEN THINK ABOUT PICKING THEIR BRAINS.
12

I'D BE HAPPY JUST TO SEE ONE AGAIN.
LET'S SET THE OTHER NETS UP OVER HERE.
YOU'LL SEE 'EM AGAIN WHEN YOU'RE NEEDED...
BUT...uh...TELL ME... WHEN YOU WERE IN THE LIBRARY... WHAT DID YOU FIND WRITTEN ABOUT 'EM?
ABSOLUTELY NOTHING.
AND NO MENTION OF MOST OF THE PLANTS AND ANIMALS ALIVE TODAY.
THE ANCIENTS MUST HAVE LIVED IN VERY BARREN TIMES.
YEAH, THE LIBRARY WAS FASCINATING! NOT MUCH WISDOM BUT LOTS OF INFORMATION.
LET'S GO BACK AND CHECK OUR OTHER NETS... MAYBE WE'VE HAD SOME LUCK...
13

NOW WATCH. YOU WANNA CATCH THE FOOT OF THE NET WITH YOUR HOOK. PULL UP EASY AND...
FLIP!
JUST KEEP YOUR BALANCE...
NOW WE MOVE DOWN THE NET... FIND THE FOOT, HOOK IT, AND...
HEY! SOMETHING'S PLAYING WITH MY...
14

MY LEG... GLUB... IT'S GOT MY... GLUB... LEG!
GIVE ME YOUR HAND...!
RRRIP
15

16

17

JACK...
SMACK
C'MON, JACK! WAKE UP!
JACK... PLEASE!
JACK... OH, JACK... IT'S TRUE... I DID TAKE THAT BOOK...
GURGLE
RETCH! CHOKE!
I KNOW THAT... WHEEZE... GIMME THE DETAILS... COUGH
IT FLOATED BY ME AS WE WERE PULLING OUT OF THE FLOODED LIBRARY. IT WAS ABOUT ATMOSPHERIC DETERIORATION IN THE DECADES BEFORE THE CATACLYSM...
AS SOON AS I HAD SNEAKED IT BACK TO MY QUARTERS I BEGAN TO READ. BUT AS I TURNED EACH SOGGY PAGE, IT WOULD DISINTEGRATE... BEFORE MY EYES! THE WATER AND FRESH AIR MUST HAVE BEEN TOO MUCH FOR IT...
YOU SEE... THERE ARE REASONS FOR KEEPING THOSE BOOKS ISOLATED...
18

IT WAS A CONFUSED, SELF-CENTERED TEXT. WHAT I COULD UNDERSTAND INDICATED THAT THE ANCIENTS WERE AWARE THAT THEY WERE THE CAUSE OF THEIR ATMOSPHERE'S DEGENERATION...
...BUT THEY WERE UNABLE TO MAKE CORRECTIONS. A STRANGE, DOOMED PEOPLE.
ANYWAY ...NOW YOU HAVE SOMETHING TO TELL YOUR COUNCIL.
MAYBE... MAYBE NOT. ALL DEPENDS.
C'MON, LET'S CHECK THAT OTHER NET... CAREFULLY. SEA SKRANKS ARE OBVIOUSLY TOUCHY ABOUT SHARING MEALS...
YOU SURE YOU'RE OK?
IF YOU CAN STAND ON THAT LEG, I THINK I CAN STAY AWAKE...
"...But we're going to take a wide detour on our way back to the ship. We have a delivery to make..."
19

WHUMP
WELL... IMAGINE MEETING YOU GOBS OUT HERE!
NOW WHY DON'T YOU HURRY AND GET THAT EVIDENCE BACK TO THE COUNCIL BEFORE IT GOES BAD.
TELL THEM THEY CAN STICK IT IN THEIR POT AND STEW IT!
HA! HA! HA! HA!
SO JACK... WHEN ARE YOU GONNA TEACH ME TO DRIVE?
THE END
20

TXAKURRA IS A FARM DOG AND THAT MEANT HE MADE DO AS A STOCK DOG, A GUARD DOG, A HUNTING DOG, AND A RETRIEVER. IN AN AGE FILLED WITH UNCERTAINTY, ADAPTABILITY COUNTS FOR A LOT.
DOG'S LIFE
Steve Stiles
EVERY MORNING THE MASTER WOULD OPEN THE PENS AND TXAKURRA, WELL-TRAINED AND SAVVY, WOULD TURN HIS FLOCK OUT TO PASTURE.
HE HERDED THEM OUT PAST THE STOCKYARD WHERE THE BEHEMOTHS GREW FAT FOR SLAUGHTER.
HE HERDED THEM UP INTO THE GRASSY HILLS. FAR BELOW, THE MASTER AND OTHERS TILLED HARSH VOLCANIC SOIL.
1

AS THE SUN BEGAN TO CLIMB, THE SHEEP SETTLED IN TO GRAZE, TXAKURRA MADE A WIDE CIRCLE AROUND THE FLOCK AND METHODICALLY RESTAKED HIS TERRITORY...
... A NICETY SOME OPPORTUNISTS IGNORED.
WHILE IT WAS STILL EARLY, A COUPLE FROM THE LOCAL PACK DRIFTED BY. THEY CAME OFTEN BUT NEVER CLOSE ENOUGH TO UNNERVE THE FLOCK. THEY WOULD NEVER CROSS TXAKURRA'S LINE... THEY UNDERSTOOD.
THEY SILENTLY CALLED TO THE FARM DOG, COAXING HIM TO THROW OFF HIS BONDS AND RUN WITH THE PACK.
IF TXAKURRA WAS MOMENTARILY TEMPTED, HIS CHARGES DID NOT NOTICE THE LAPSE.
2

IN THE HEAT OF MIDDAY, MOST PREDATORS DISAPPEAR INTO THE COOL OF THEIR DENS.

THEN SOMETIMES THE SHEEP WOULD GROW LESS CAUTIOUS... EVEN **BOLD**... AND BEGIN TO STRAY.

TXAKURRA COULD REMEDY THIS WITHOUT MOVING. HE MERELY **PRETENDED** SOME MENACE WAS NEAR...

... AND THE SHEEP WOULD COME BACK TOGETHER. IT WAS A GOOD JOKE **AND** EFFECTIVE. THE SHEEP NEVER CAUGHT ON.

TWEEEE
SO, TXAKURRA! HOW WAS WORK TODAY?
THE FLOCK DELIVERED BACK TO SAFEKEEPING, HIS BELLY FULL, TXAKURRA BEDDED DOWN FOR THE NIGHT.
4

THIS WAS THE MOST UNEASY TIME FOR A DOMESTIC IN SUCH CLOSE DAILY CONTACT WITH THE WILD.
THEY CAME AGAIN SHORTLY AFTER MOONRISE.
THEIR SCENT AND THE WEIGHT OF THEIR PRESENCE PRESSED DOWN ON THE FARM DOG.
IN THIS HOUR, WHEN HIS DUTIES WERE CONCLUDED FOR THE DAY, THE HOLES IN HIS LOYALTY WERE FILLED WITH LONGING FOR THE PACK.
INSTINCTS HELD IN ABEYANCE BEGAN TO WELL UP.
TONIGHT THE WILD CALL MIGHT HAVE BURST HIS CHAINS OF SERVITUDE...
...EXCEPT FOR ONE THING...
5

WITH ALL THE STEALTH AND CUNNING OF HIS KIND THE CUTTER HAD GLIDED INTO THE FARMYARD DOWNWIND FROM THE DOG AND THE FLOCK.
BAAA BAAA
CONFUSED FOR ONLY A SPLIT SECOND, THE DOG CHARGED THE INTRUDER.
NEVER BEFORE HAD THE WARY CUTTER VENTURED THIS CLOSE TO THE HAUNTS OF MEN. HUNTING IN THE HIGH PLAINS HAD BEEN VERY POOR.
SSSSS
SSSS
ALL MEMORY AND LONGING FOR THE PACK MELTED FROM TXAKURRA AS HE CONFRONTED THIS DANGER TO HIS FLOCK.
BAA BAA
BAA
BAA
BAA
BAA
6
HE SOUNDED THE ALARM AND GRIMLY DUG IN TO DIVERT THE BEAST UNTIL HELP ARRIVED.
RRF! ARF! ARF! ARF!

BAA
BAA BAA
BAA BAA
BAA
REEEEEEE
SNAP!
SNAP!
YOWL!
BAA BAA
BAA
BAA
BAAA
BAA
7

CRACK!
CRACK!
ZZIING
ZZZIING
BWEE
DAMN! THEY'RE GETTIN' BOLD!
POW!
HOLD ON, BOY! LEAVE 'IM GO AND COUNT YERSELF LUCKY!
HEY! LOOK AT THAT CHUNK OUTTA HIS HIDE!
LATER THAT NIGHT THE PACK RETURNED...
...BUT NOW THEIR SPHERE OF INFLUENCE HAD SHRUNK CONSIDERABLY.
8
TXAKURRA QUIETLY NOTED THEIR DISTANT PRESENCE. TOWARDS DAWN HE CAUGHT A LITTLE SLEEP BEFORE STARTING A NEW DAY.
END

JACK...
HANNAH! GOOD LORD!
WHERE HAVE YOU BEEN?!
FOUNDLING
THAT INVOLVES QUITE A STORY...
...HAVE A FEW MINUTES?
MARK SCHULTZ © 1988
1

"You do? Good, 'cause this should interest you.
"Remember this morning?... Seems so long ago... We were in the city and you were having another of your little squabbles with the governors..."
"Until she came and interrupted. Maia Abrelatas pleaded with you... begged the governors to reconsider their refusal to renew the search for her son..."
HANNAH DUNDEE! YOU'RE A SKILLED TRACKER, AREN'T YOU?! YOU MAY BE MY LAST HOPE...
PLEASE... LISTEN TO MY STORY... OF HOW MY SON WAS STOLEN FROM ME TEN YEARS AGO!
"And I'm sure you remember how the governors turned her away... while you looked on!
"But then, in desperation, she cornered me in the hall..."
HE WAS ONLY THREE THE NIGHT HE WANDERED OUT THE BACK OF MY FARMHOUSE. IT WAS ONLY MINUTES BEFORE WE REALIZED HE WAS GONE...
BUT IN THAT SHORT TIME THE HYENAS HAD SILENTLY COME AND GONE...
WE FOUND THEIR SPOOR ALL AROUND HIS... AND BLOOD...
2

FOR DAYS WE COMBED THE WILDS AND FOUND NOTHING.
THEN THEY CALLED THE SEARCH OFF... ALTHOUGH I KNEW HE WAS STILL ALIVE! I COULD FEEL HIM OUT THERE.
FOR TEN YEARS I'VE PETITIONED THE GOVERNORS... AND TENREC... TO RENEW THE SEARCH.
EVERYONE THINKS I'M CRAZY, BUT LISTEN... TWO WEEKS AGO MY SON REAPPEARED... AT MY WINDOW!
NOW HE COMES TO THE FARM NIGHTLY... AND AS SOON AS HE REALIZES I'VE SEEN HIM, HE MELTS AWAY... LIKE A PHANTOM.
HE'S TEN YEARS OLDER... BUT I KNOW MY SON AND SOMETHING IS KEEPING HIM FROM RETURNING TO ME! PLEASE..., HELP ME!
"It was a mad tale... but I was moved by her very real sadness.
"So I did what you should have done, Jack. I went with her back to the farms..."
"There were many tracks left by an adolescent male leading to and from her window... Some were very recent..."
"It didn't make any sense... How could a child remain alive for ten years in the interior?!
"Still, someone had made those tracks, and he couldn't be far away."
"I decided to follow the spoor..."
3

"It wasn't easy.
"You know I'm not bragging when I say that very few could have kept to that trail."
"Then the trail freshened... I knew he must be close by..."
"By then it was late in the afternoon and I was miles into the jungle."
"I was about to give it up when suddenly... there he was!"
"He was totally absorbed in his reflection... I got quite close to him before..."
"...He started up like a frightened shilliwak...
"He looked like he'd never seen a human before. But he didn't run... although I think he wanted to..."
4

"He allowed me to calm him... He seemed fascinated by my voice..."
IT'S OK... I'M NOT GOING TO HURT YOU...
"I could tell he understood my words..."
WHO ARE YOU? WHAT'S YOUR NAME? HOW DID YOU GET HERE?
"Then, without a sound, he picked up a sharp stone and..."
HYENAS DRAG ME HERE
"There was something about his figures... with their peculiar frames... that tugged at my memory..."
"Something familiar, yet alien..."
"I don't believe he knew how to speak. He certainly didn't mind scrawling out replies to my questions, although most of 'em didn't make any sense...
"He couldn't tell me where he came from, or who taught him to write..."
5

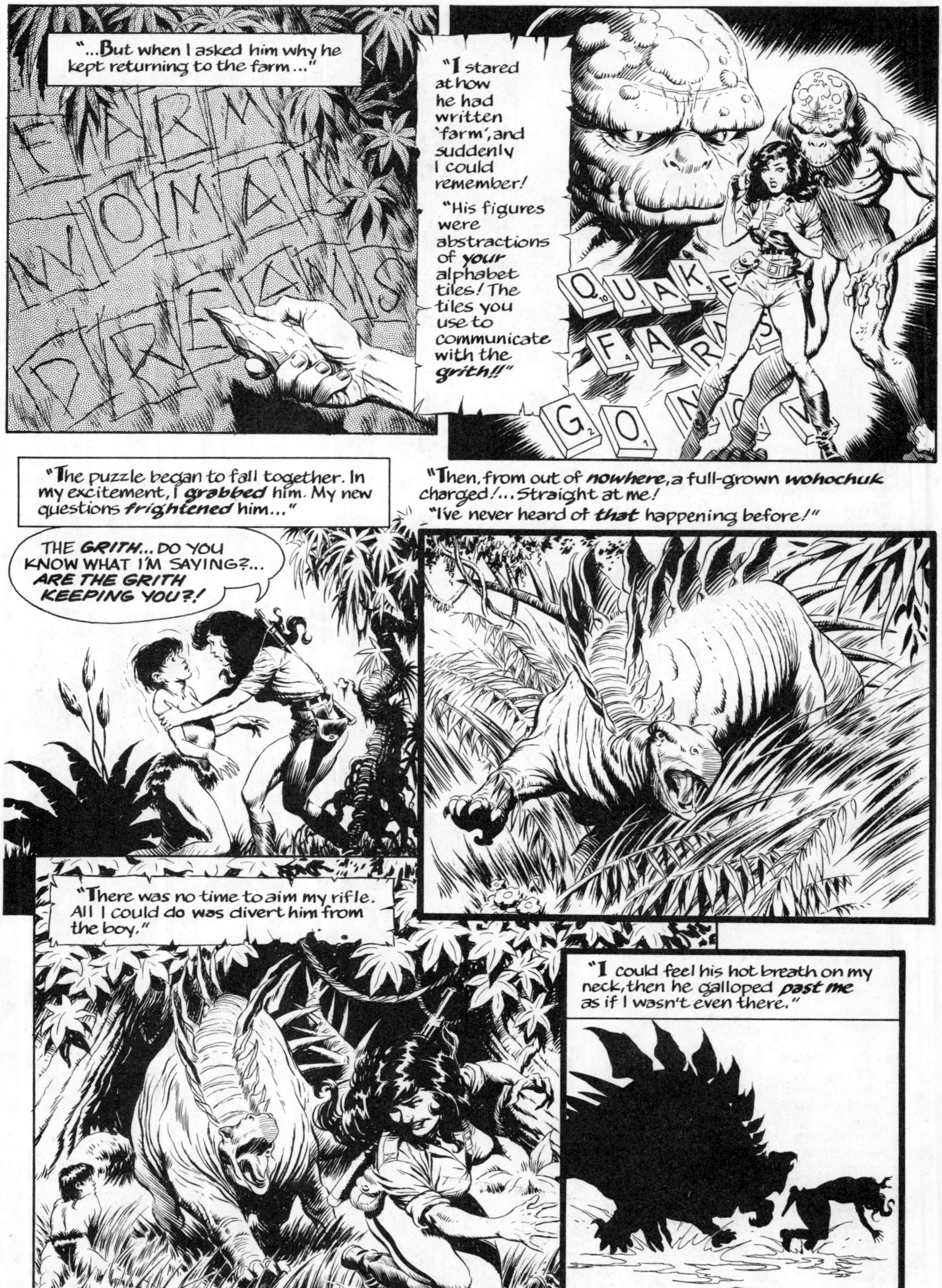
"...But when I asked him why he kept returning to the farm..."
FARM WOMAN DREAMS
"I stared at how he had written 'farm', and suddenly I could remember!
"His figures were abstractions of *your* alphabet tiles! The tiles you use to communicate with the *grith!!*"
QUAKE FARMS GO NOW
"The puzzle began to fall together. In my excitement, I *grabbed* him. My new questions *frightened* him..."
THE *GRITH*... DO YOU KNOW WHAT I'M SAYING?... *ARE THE GRITH KEEPING YOU?!*
"Then, from out of *nowhere*, a full-grown *wohochuk* charged!... Straight at me!
"I've never heard of *that* happening before!"
"There was no time to aim my rifle. All I could do was divert him from the boy."
"I could feel his hot breath on my neck, then he galloped *past me* as if I wasn't even there."

"The boy was gone..."
"But he had left a trail that I was determined not to lose until I came face to face with a grith!"
"Even so, I would have lost the spoor in the failing light if I hadn't seen their distant figures."
"I tore after them. I was so close..."
"When I crested the hill, there was no sign of them. But the spoor I found outside a cave confirmed all my suspicions..."
7

"I rigged a makeshift torch...and driven by unreasoning obsession..."
"I descended into the domain of the grith!"
"It couldn't have been long before I became aware of the eyes... pinning me down... helpless... defenseless..."
"The world became a blur... I forgot the boy... I forgot the grith... I forgot everything in a blind panic!"
"I had to get out! I had to get away!
"I can't remember how far I ran before I began to wonder just what had frightened me."
"And so I stumbled back here..."
8

BUT YOU ALREADY KNOW ALL ABOUT THE BOY, DON'T YOU?!
YOUR FRIENDS THE GRITH MUST HAVE CLUED YOU IN!
THE GRITH FOUND THE BOY TORN APART BY THE PACK... THEY SAVED HIS LIFE... THEY DIDN'T HAVE TO, YOU KNOW. THEY'VE... CARED FOR HIM.
NOW THEY'RE IN A BAD SPOT... THE BOY CAN'T GO HOME... NOT AFTER WHAT HE'S SEEN AND LEARNED. HE EVEN THINKS LIKE A GRITH!
BUT THE FACT IS, HE'S ALSO HUMAN. AND HE'S AT THAT AGE NOW. HE'S VERY CONFUSED, AND IT'S BECOMING A PROBLEM.
I THINK THEY'LL BE TAKING HIM AWAY FROM HERE.
AND WHAT DO I TELL HIS MOTHER?
THE SAME THING I'VE BEEN TELLING HER.
I CAME VERY CLOSE TO MAKING CONTACT WITH THE GRITH TODAY, JACK. I RISKED MY LIFE TO DO IT... BUT THEY TURNED ME AWAY.
BUT THAT'S NOT ALL THAT'S BOTHERING YOU, IS IT?
WHY? THEY KNOW I WANT SOME ANSWERS! I'M TIRED OF BEING JERKED AROUND.
RISKED YOUR LIFE? DO YOU REALLY THINK YOU COULD HAVE MADE IT IN ONE PIECE THROUGH THE INTERIOR AFTER DARK... UNLESS THE GRITH MADE DAMN SURE YOU DID?
AND I WOULDN'T WISH FOR SOMETHING THAT MIGHT JUST HAPPEN.
BY THE WAY, HAVE YOU HAD DINNER YET?
THE END
9

UHHH... EASY... EASY... DON'T SCREW UP NOW, YOU IDIOT... THEY'RE COUNTIN' ON YOU... GET A HOLD OF YERSELF...
INTRUSION
THINK ABOUT WHAT YER DOIN'... REMEMBER WHY YER HERE...
TENREC... YA FILTHY SKANK! ...YA MURDERED 'EM BOTH... WRENCH... HAMMER... MY BROTHERS! NOW I'M GONNA MAKE EVERYTHING EVEN UP! NOW YER GONNA PAY...
...GOTTA AVENGE THE TERHUNE FAMILY HONOR... YEAH I'M COMING FOR YA, TENREC!
Steve Stiles

MAN... IT'S LIKE A WIND TUNNEL IN HERE!
WHEN ARE WE GONNA REPLACE THOSE PANES, JACK? I SWEAR, THIS OLD PILE AINT NOTHIN' BUT A SIEVE!
Y'KNOW, IF ANYBODY WANTED TO GET IN HERE WITHOUT US KNOWIN'...
OH, I HARDLY THINK THAT'S ANYTHING TO WORRY ABOUT. REMEMBER, KIRGO -- IT WAS THE TORNADO THAT BLEW OUT THE WINDOWS.
NO, IT'S NOT MEN BREAKING IN HERE THAT CONCERNS ME...
IT'S NATURE... ALWAYS BATTERING AT THE WALLS, PATIENTLY CLAWING AT THE FOUNDATIONS... THE EARTH IS GONNA BE THE DEATH OF THIS GARAGE, NATURE IS THE WOLF AT THE DOOR...
...THREE WEEKS... IT TOOK ME THREE WEEKS OF SCRAMBLIN' AN' CRAWLIN' ALL OVER THESE DAMN CLIFFS TO FIND A CRACK IN THE FOUNDATION.
!
2

WELL, I'LL ADMIT THE WEATHER **IS** CERTAINLY DOIN' ITS PART TO CHEW THIS PLACE APART...
... PASS ME THAT RATCHET...
REMEMBER THAT SANDSTORM A COUPLE OF YEARS BACK? MUST'VE BLASTED A **FOOT** OFF THE OUT-SIDE, AND IT TOOK A MONTH TO EXCAVATE THE INSIDE.
PUFF, PUFF! STUPID DAMN SLITHER! SHOULDA KNOWN I'D RUN INTA ONE SOONER OR LATER...
...TORE ME UP GOOD... LOSIN' BLOOD...
NOW WHAT?!
RRRUU UMMMMB-BBLLLEEE
3

THAT'S EXACTLY WHAT I MEAN... THE EARTH AND SKY ARE CONSPIRING TO ERODE THIS...
HEY! EARTHQUAKE!
RRRRRUUUMMBBLLLEE
WHEW! GUESS THAT'S ALL...
JUST A MINOR SHOCK... HANNAH TOLD ME SOMETHIN' ABOUT PLATES DOWN IN THE GROUND...
...SHE SAYS THEY'RE ALWAYS BUMPIN' UP AGAINST EACH OTHER...
GET UP... KEEP MOVIN...
...WRENCH... HAMMER... I'M DOIN' THIS FOR YOUSE...
...ALMOST THERE...
...TOOK ME AN ENTIRE WEEK TO LOOSEN THESE STONES...
THE GEOTHERMAL FURNACES! HERE ON THE LOWEST LEVEL, THE HEART OF TENREC'S GARAGE!
A FEW WELL-PLACED EXPLOSIVES AND THEN A CHAIN REACTION THAT'LL BLOW THE ENTIRE PLACE TO HELL AND BACK!
4

...AND HOW ABOUT THOSE LIGHTNING STORMS? THEY'RE TAKIN' THEIR TOLL...
SO YOU SEE, IT'S HARD FOR ME TO GET WORKED UP ABOUT HUMAN ENCROACHMENT WHEN I'VE GOT TORNADOES, QUAKES, SANDSTORMS, AND ELECTRICAL STORMS TRYING TO GRIND THIS GARAGE OFF THE FACE OF THE EARTH.
OKAY, LET'S START 'ER UP...
...HURT BAD... GOT TO RIG THIS UP AND GET OUT BEFORE I PASS OUT...
OH, NO!!
OH, NO!!
RRRRUU UMMMMBRBLLLEEE
5

RRRUUMMMMBBBLLE
CREAK!
UH!
TSSSSSS
SSSSSSSSS
THUD!
SSSSSSSSSSSSSSSSSSSS
POOM!
POOM!
POOM!
POOM!
POOM!

RRRRUUMMBBLLEE
THAT FELT LIKE THE REAL THING...
I DON'T THINK THE GARAGE CAN TAKE MUCH MORE. MAYBE THAT WAS OUR EVICTION NOTICE...
C'MON... WE BETTER CHECK THE GEOTHERMAL WORKS. IF THE FURNACES HAVE BEEN DAMAGED WE COULD BE IN BIG TROUBLE!
EXIT
THE DAMAGE CAN'T BE THAT BAD OR WE WOULDN'T BE HERE ANYMORE...
LET'S JUST WAIT AND SEE...
SURE HOPE WE DON'T GET AN AFTERSHOCK NOW...
7

!
ONLY MINOR STRUCTURAL DAMAGE...
BUT FOR CRYIN' OUT LOUD! HOW'RE WE EVER GONNA CLEAN THIS UP?!
JUST WHEN I THOUGHT WE'D SEEN EVERYTHING THIS LAND COULD THROW AT US...
WAIT A MINUTE! THE QUAKE DIDN'T BURST THE SLAG PIPE! THE VALVE'S OPEN!
SOMEONE LEFT THE VALVE OPEN!
WHO WOULD'VE DONE THAT?
I DUNNO, BUT IF THAT SLAG HADN'T BEEN LEAKED, AND IF THE QUAKE HAD SHAKEN THE SLAG BACK INTO THE FURNACE, THE PRESSURE WOULD'VE SENT THE WHOLE GARAGE UP LIKE A VOLCANO!
8
WE'RE JUST LUCKY SOMEONE DID OPEN THAT VALVE...
LUCKY?
THIS ISN'T WHAT I WOULD CALL LUCK...
END

1

GREEN AIR
NICE DAY FOR SOARING, REMFRO!
EXCELLENT DAY FOR SOARING!
SURE, JACK.
ALL SAILPLANES MUST MOVE FORWARD TO STAY ALOFT. ONCE I'M UP, GRAVITY WILL BEGIN PULLING ME IN A SHALLOW DOWNHILL COAST--ONE FOOT DOWN FOR EVERY 23 FORWARD...
TO STAY UP, I NEED LIFT... GREEN AIR. THE SUN WARMS THE EARTH, WHICH IN TURN THROWS COLUMNS OF HEATED AIR TO THE SKY.
THESE THERMAL UPDRAFTS WILL PUSH ME BACK UPSTAIRS.
WHY DON'T YOU TELL ME AGAIN HOW THIS IS SUPPOSED TO WORK...
WELL, I DON'T KNOW HOW YOU CAN BE SO SURE OF SOMETHIN' YOU CAN'T SEE...
AND I WON'T PROMISE YOUR PLANE WILL EVEN GET OFF THE GROUND...
OH, SHE WILL!
YOUR RESTORATION IS FLAWLESS, JACK. YOU'VE EXECUTED ALL MY SPECIFICATIONS PRECISELY!
MARK SCHULTZ 1988 ©
THANKS AND A TIP OF THE AILERONS TO VINCE RUSH, AERONAUTICS ADVISOR.
2

NOW LET'S GET THIS SHOW ON THE ROAD! MY COMPATRIOTS ARE WAITING!
HEY, DON'T FORGET YOUR NOTES... JUST IN CASE...
HANNAH, MY DEAR... UP THERE I WON'T HAVE TIME TO CONSULT MY NOTES...
TAKE IT ANYWAY.
I DIDN'T RUIN MY EYES RECOPYING THIS THING'S CRUMBLING FLIGHT MANUAL SO YOU COULD IGNORE IT...
...YOU DON'T KNOW WHAT MIGHT HAPPEN UP THERE.
THE ZEKES WILL BE MY GUIDES.
I'VE BEEN PREPARING FOR THIS ALL MY LIFE... YOU KNOW HOW I'VE STUDIED THE ZEKES... THEY'RE GLIDERS, TOO... PERFECT GLIDERS... THEY'LL TEACH ME TO SOAR... THEY'LL LEAD ME TO THE GREEN AIR THAT KEEPS US ALL ALOFT...
"This is an historic day, Hannah. Not since before the great cataclysm has man soared with the reptiles!"
GO!
GO!
CHECK!
VVVRROOOM!
3

REMEMBER TO STAY OVER THE LAND!
...AND BE CAREFUL!
NOSE UP... GAIN ALTITUDE... MORE STICK, MORE STICK...
TOO MUCH STICK! TOO STEEP! DON'T STALL, DON'T STALL!
OH, NO!
OH, NOOOOOOO!
TWANG!
!
4

CAST OFF, YOU IDIOT, BEFORE YOU TEAR 'ER APART!
WHAT THE HELL...?
HERE GOES NUTHIN'...
PING!
ATTABOY, REMFRO!
NOW TRIM 'ER... EASY DOES IT...
I DON'T BELIEVE IT... WOULD YOU LOOK AT THAT...

I'M DOIN' IT. I'M REALLY DOIN' IT!
A LITTLE LEFT STICK LEFT RUDDER AND I'LL JOIN THE FLOCK...
WHOA!
GENTLY, GENTLY! WATCH THE ZEKES, SEE HOW EASY THEY DO IT...
OOOOOH! EASE UP, REMFRO! DON'T MUSCLE 'ER!
HE'S FIGHTING THE CONTROLS...
HOW DO YOU COME TO BE SUCH AN EXPERT?
YOU DON'T THINK I'D ALLOW HIM TO GO UP IF I HADN'T THOROUGHLY STUDIED THIS BUSINESS MYSELF?
WANNA LEARN?
NO THANKS! YOU'LL NEVER CATCH ME UP THERE WITHOUT AN ENGINE...
MAYBE SOMEDAY I'LL RECONDITION ONE OF THOSE POWERED BABIES AND...
WHOOOOPS!
6

I'M LOSING ALTITUDE FAST...
YES! WHAT A SURGE!
LIKE GETTING A KICK IN THE PANTS!
LET'S SEE... THOSE ZEKES OVER THERE... SPIRALING UP... HAS TO BE A THERMAL.
DO I HAVE ENOUGH ALTITUDE LEFT TO REACH IT?
AND I WAS RIGHT! THE ZEKES SEEK THE SAME GREEN AIR THAT KEEPS ME GOING...
NOW THAT I'VE GOT THIS LICKED, I CAN STAY UP ALL DAY!
HEY! WHERE'S EVERYBODY GOING?!
7

THEY'RE ALL HEADED OUT TO SEA!
WELL, OF COURSE! WHY NOT? THEY'VE GAINED ENOUGH ALTITUDE FROM THE THERMAL TO ALLOW THEM TO GLIDE OUT TO THE FISHING BANKS...
THEY'LL CATCH OTHER THERMALS OUT THERE.
I'M ONTO YOUR GAME, MY WINGED FRIENDS! I'M COMIN' TOO!
YES, THE SKY'S NO LIMIT ONCE YOU'VE LEARNED HER SECRETS!
WE FEARLESS DENIZENS OF THE AIR WORRY ABOUT TOMORROW WHEN IT COMES!
FORGET THE CAR, JACK. WE NEED A BOAT.
I THINK OUR BIRDMAN MUST BE SUFFERING FROM OXYGEN DEPRIVATION...
?
8

KIND OF SCARY, ACTUALLY. WE'RE PRETTY FAR OFFSHORE...
THEY'RE DIVING... MUST SEE A SCHOOL DOWN THERE.
GOOD FISHING!
I WISH THEY'D FINISH SOON. I'M GETTIN' AWFUL LOW AND THOSE WAVES ARE LOOKIN' AWFUL BIG.
C'MON, GUYS! ISN'T ANYONE INTERESTED IN MAKING IT BACK TO SHORE?
I TRUSTED YOU!
WHERE ARE THE THERMALS?!
9

WAIT A MINUTE, GUYS! WHAT'S THIS? DON'T LEAVE ME! I CAN'T KEEP UP!
WE'VE GOT TO FIND A THERMAL! THERE'S NO WAY WE CAN STAY UP ANY LONGER... UNLESS... UNLESS... CHOKE
...UNLESS YOU'RE A ZEKE AND CAN FLAP YOUR WINGS...
...AND FLY BACK... SOB
10

THERE... HE'S DOWN!
ALL THE TIME...ALL THE WORK I PUT INTO THAT PLANE.
I HOPE YOU WATERPROOFED HER.
SIGH
JUST THINK, JACK! TODAY WE ACCOMPLISHED THE FIRST MANNED FLIGHT IN HUNDREDS OF YEARS!
THE XENOZOIC SKIES WILL NEVER BE THE SAME.
I DON'T KNOW ABOUT THAT, HANNAH, BUT THERE MUST BE AN EASIER WAY TO FEED THE SHARKS...
...ALTHOUGH THEY DON'T SEEM TO BE COMPLAINING... HEH, HEH!
THE END
11

FOUL WEATHER
ANGRY WAVES LASHED THE LONELY TOWER, LONG ABANDONED TO THE CONTINUALLY RISING SEA. THE GROWING POWER OF THOSE WAVES, COUPLED WITH DARK, HEAVY CLOUDS PILING UP IN THE SOUTHERN SKY, HERALDED THE COMING OF A STORM OF UNUSUAL FEROCITY...
FROM HER PERCH WITHIN THE TOWER, MIKLA VAN ERMINE APPREHENSIVELY SCANNED THE CHOPPY STRAIT BETWEEN HER AND THE DISTANT CITY.
THEN SHE SAW HIM, A MOTORIZED SPECK SLAPPING ACROSS THE WHITECAPS. VICE TERHUNE WAS COMING TO MEET HER AT THIS CRUMBLING EDIFICE, THIS RENDEZVOUS FOR PIRATES, BLACK MARKETEERS... AND THIEVES.
steve stiles

VICE, BABY! DID YOU PULL IT OFF? WHERE ARE THE OTHERS?
DEAD.
WE MADE IT INTO THE CITY'S TREASURE VAULTS NO PROBLEM, BUT AFTER THE HEIST KLEGG TRIPPED AN ALARM. THE *@#!! GUARDS GOT HIM AN' SLY. I BARELY MADE IT TO THE BOAT!
SO WHAT'S THE LOOT?
I DUNNO.
YOU DON'T KNOW?!
IT'S IN A STRONGBOX. I BEEN TOO BUSY SAVIN' MY SKIN TO LOOK INSIDE.
BUT IT'S GOTTA BE WORTH PLENTY OR THE ANCIENTS WOULDN'T HAVE KEPT IT SO CAREFULLY HIDDEN.
WELL, LET'S OPEN IT!
UH,UH... NO TIME, BABE. JACK *!!! CADILLAC'S GOTTA BE ON MY TAIL BY NOW!
AND EVERY SECOND I WASTE HERE IS BRINGIN' HIM CLOSER...
I GOTTA TAKE THE LOOT AND MAKE MYSELF SCARCE!
BUT, VICE! THERE'S A BAD STORM BREWING! YOU CAN'T RISK THE SEA NOW!
WE COULD HOLD AN ARMY OFF HERE!
AN ARMY, MAYBE. BUT NOT !!@*!! TENREC. HE'S ALREADY WIPED OUT HALF MY FAMILY. BESIDES NOW HE KNOWS WHO STOLE HIS SPEEDBOAT!
I'LL RISK THE STORM!
THEN AT LEAST HIDE THE LOOT HERE!
2

HA, HA! NO WAY, BABE!
THE BOX STAYS WITH ME! NOT THAT I DON'T TRUST YOU! HA, HA, HA!
MEET YOU BACK HERE ONCE THINGS COOL DOWN!
?
BUMP!
WHO...?
JACK TENREC! WHAT THE HELL?! YOU'VE BEEN HERE ALL ALONG!
DROP YOUR GUN... SLOWLY.
AND THAT'S THE STRONGBOX, ISN'T IT?! YOU SWIPED THE LOOT AND LET VICE SAIL OFF INTO THE STORM...
IT'S EVEN WORSE THAN THAT! I HAPPEN TO KNOW THAT THIS IS A FULL FLEDGED HURRICANE MOVIN' UP THE COAST.
I WONDER IF VICE WOULD'VE GONE BACK OUT IF HE'D KNOWN THAT?
YOU BILGE RAT! HE'S GONNA DIE OUT THERE!
THAT'S WHAT I HOPE!
DANDY LITTLE PLAN, HUH?
GET INSIDE!
I'M NOT GONNA STAND HERE GETTING WET WHILE I DECIDE HOW TO DO AWAY WITH YOU...
3

HEY, DON'T GET SO UPSET! THIS BOX IS WHAT YOU REALLY WANT, ISN'T IT?
C'MON, MIKLA! WE BOTH KNOW YOU'VE BEEN USING VICE!
THIS IS WHAT YOU WANT! YOU SHOULD THANK ME FOR SWIPIN' IT FOR YOU!
DON'T YOU WANT TO SEE WHAT'S INSIDE?
WHY DON'T YOU TAKE A LOOK?
IT'S ALL YOURS NOW!
OK... PUT IT DOWN... EASY... NO TRICKS...

THUNK!
5

MIKLA!
SO, YOU BEEN WORKIN' WITH *!!@ø* TENREC! ...YOU TWO TIMIN'...
VICE!
NO, VICE! IT'S NOT LIKE YOU THINK! TENREC...
SHADDUP! I CAN THINK FOR MYSELF AND I THINK YOU TWO BEEN SCHEMIN' BEHIND MY BACK!
DIDN'T COUNT ON THE STORM BLOWIN' ME BACK HERE, DID YOU?
WELL, YOU'VE GOT EVERYTHING ALL FIGURED OUT, DON'T YOU?
@*!!#@!! DON'T LISTEN TO HIM, VICE!
DRILL HIM!
OH, I WILL, BABY!
CLICK!
...BUT FIRST...
BANG!
6

LOOK!

PAPER... PAPER! SO THAT'S THE TREASURE VICE TERHUNE RISKED AND LOST HIS LIFE FOR! I GUESS THE ANCIENTS HAD A WEIRD SENSE OF HUMOR!
WELL, KEEP IT, MIKLA! BE MY GUEST! IT'S ALL YOURS!
HOPE YOU CAN FIND SOME USE FOR IT!
8
END

THE GROWING POOL
THIS IS A STORY ABOUT LIFE AND MYSTERY AND FACTS...
HARD FACTS...AS BRIGHT AND PELLUCID AS A DIAMOND, BUT AS SLIPPERY AND HARD TO GRASP AS A MOSSY RIVERSTONE...
...AND ASSUMED FACTS... DEVIOUS AND SUBVERSIVE ...OFFERING NO RESISTANCE AS THEY SOFTLY ENGULF YOU.
NO ONE HOLDS A MONOPOLY ON THE FACTS.
AND THAT'S A FACT.
BEFORE WE BEGIN, LET'S SLIP BACK FOR A MOMENT...BEFORE THE CONTINENTS BUCKLED AND ERUPTED...
BACK DURING THE EARLY DISASTERS, WHEN THE POISONED ATMOSPHERE WENT BERSERK, AND THE SEAS ROSE...LET'S SLIP BACK IN TIME TO A LONG SUBMERGED ISLAND NATION...
MARK SCHULTZ © 1988
THANX TO LESLEIGH LUTTRELL, WHOSE SAGE ADVICE IS HEREIN TWISTED AND MISUSED.
1

JAPAN, 2010!
THE ARCHEOPLASM... HAS IT BEEN SHIPPED YET?
THE COMPANY COURIER JUST LEFT...
CARE TO WATCH?
I SUPPOSE SO... I SUPPOSE I SHOULD WANT TO SEE THE SECRET OF THE ELIXIR VITAE... WHICH WE WRESTLED INTO EXISTENCE AFTER SO MANY LONG, HARD YEARS... SIMPLY HANDED OVER TO OUR ENEMIES.
DON'T BE SO BITTER, HIDEYORI. IT'S THE WAY OF THE WORLD... WE HAVE SOMETHING THEY NEED... THEY HAVE THE MILITARY MIGHT... WE HAD NO CHOICE.
BESIDES... THE POINT IS MOOT. THE ARCHEOPLASM WILL DO THEM NO GOOD NOW... THEY'RE TOO FAR GONE.
I HEAR THEY'RE GOING TO SUSPEND IT... BURY IT IN BEDROCK FOR FUTURE USE... IF THE EARTH EVER SETTLES DOWN AGAIN.
A PRACTICAL IF COLD-BLOODED DECISION.
AND THE SHOE COULD HAVE BEEN ON THE OTHER FOOT... THEY VERY EASILY MIGHT HAVE BEEN THE ONES LUCKY ENOUGH TO BE VISITED BY THE METEORITE THAT UNLOCKED THE SECRET.
THERE IT GOES...
RIGHT OVER THE TRILOBITE BEDS THAT THE ARCHEOPLASM MADE POSSIBLE...
2

Word of the meteor's impact, and the strange vault it unearthed, eventually sifted into the City in the Sea. I was assigned by the council to assemble a field unit to investigate.
The governors' policy has always been to keep the tribe's contact with pre-cataclysmic technology to a minimum, at least until an evaluation can be made. News of this discovery and of my expedition was accordingly down played.
Some persons, however, are not so easily deterred.
THAT'S THE WASSOON...
HANNAH DUNDEE...
What my team and I found was a multi-chambered vault of mid-20th century design. Apparently hermetic, and undamaged until impacted, it lay under 15 feet of bedrock. The outermost portal had been ruptured.
The meteorite had disintegrated upon impact.
DR. BULGAR, I PRESUME?
WHAT'S YOUR BUSINESS HERE?
SAME AS YOURS...THE INCREASE AND DIFFUSION OF KNOWLEDGE AND ALL THAT.
CARE TO SHOW ME WHAT YOU'VE GOT HERE?
A REPOSITORY OF SOME SORT... IT DATES FROM THE MID-20th CENTURY.
NO... THEY HADN'T DEVELOPED CERAMICS LIKE THESE UNTIL THE EARLY 21st...
I had to admit she was probably right. Dundee's familiarity with the past was impressive...
4

So, as she was obviously not going to just go away, and as her reputation prevented any of us from trying to *forcibly* remove her, I decided to put the Wassoon's knowledge to good use.
The next day we forced open the inner portal.
But even as we broke into that inner sanctum, *trouble* was creeping up on us. The last man in the world I wanted sticking his nose in our operation was on his way...
In the center of the chamber we found a cache. Its integrity had not been disrupted.
It took us another day before we were sure we could open the cache without damaging it...
UGH... LOOKS LIKE THIS WENT BAD...
The cache held something... *organic*.
A PLASMAGEL? AND THAT *LOOKS* LIKE ALGAE AND MOLD...
5

That's when all the trouble started.
I GUESS I'M A LITTLE TOO LATE...
THAT SHOULD NOT HAVE BEEN OPENED UNTIL YOU HAD IT SAFE IN THE CITY.
TENREC! I THOUGHT YOU WERE AWAY... FAR, FAR AWAY...
I WAS... BUT WORD GETS AROUND.
THOUGHT YOU COULD AVOID MY INPUT, HUH?
HELLO, HANNAH.
CAN'T SAY I'M SURPRISED TO SEE YOU HERE...
SEE HERE, TENREC! EVERYTHING'S UNDER CONTROL. YOU CAN'T BULLY...
OH, CAN'T I? JUST TRY AND STOP ME!
HERE IN THE INTERIOR, MY WORD GOES! AND MY RULE HAS ALWAYS BEEN WE DON'T SCREW WITH PRE-CATACLYSMIC DISCOVERIES UNTIL WE GET 'EM BACK IN THE CITY WHERE ANY POTENTIAL ENVIRONMENTAL HARM WILL BE MINIMALIZ...
MY GOD!
LOOK!
GOOD LORD!
6

WHAT THE HELL IS THAT?!
STOP IT! BEFORE IT GETS AWAY!
WAIT! DON'T TOUCH IT!
GIMME YOUR GLOVES...
WHAT...?
The goop flowed through Tenrec's ineffective grasp, and slithered down a crack in the floor...
WHERE WOULD IT BE GOING?
W-WATER, MAYBE? IT WOULD NEED MOISTURE...
...MAYBE THE WATERING HOLE!
We made it to the pool just as the ambulatory slime slipped under the surface.
For a few moments we stood in shocked disbelief, unable to fathom what our eyes had told us was true...
7

OH, THIS IS JUST GREAT! WHO KNOWS WHAT ABOMINATION YOU JUST LET LOOSE!
MAYBE A PLAGUE, HUH?
OR MAYBE SOMETHING THAT'LL JUST EAT ITS WAY THROUGH THE COUNTRYSIDE?
WELL, I GUESS WE'RE STUCK WITH IT! BUT I'M GIVING THE ORDERS NOW. I WANT THIS POOL COMPLETELY ISOLATED... NO ONE AND NOTHING IN OR OUT... AND NO ARGUMENTS!
YES, OF COURSE! A CONTROLLED EXPERIMENT!
I DON'T CARE WHAT YOU CALL IT... NO ARGUMENTS!
Tenrec may not have known what a controlled experiment was, but his methods of enforcement were sound.
He had a ring of fire burned around the pool and camp.
Guards were stationed along the scorched perimeter and at the pool, which up till now had been a watering hole for the local wildlife.
No one was allowed to leave camp. We cautiously watched for any signs of infection, and water samples were taken hourly from the pool.
8

The changes began occurring immediately...
AMAZING... EVEN MORE SPECIES IN EVIDENCE NOW. BACTERIA, ALGAE, ROTIFERS, PLANARIA, EVEN PRIMITIVE CRUSTACEA... EVERY SAMPLE I TAKE YIELDS MORE HIGHLY EVOLVED LIFE FORMS.
A COMPLETE NEW ECOSYSTEM EVOLVED IN A MATTER OF DAYS. BUT NOTHING ALIEN... NO FORMS THAT DIDN'T FIT...
The fire and smoke kept most of the wildlife from the water hole...
But every so often there'd be a problem...
BLAM!
At times like these I was almost glad for Tenrec's presence.
9

As the long days of watching and waiting wore on, Tenrec and Dundee worked tirelessly to keep the guards alert.
Even they needed the occasional break, however...

The fool guard at the north post had fallen asleep and let his fires burn out...
Eager for a chance to get to the watering hole, the cutters had made short work of him.
Jack traced the pack's spoor until he was satisfied none of the slithers had made it to the water.
Meanwhile, our ongoing analysis of the pool water showed that the last few days' explosion of life forms had ended.
Apparently, all available ecological niches had been occupied by life forms as highly developed as present environmental conditions would allow.
We theorized we would see no further evolutionary development until conditions in the pool changed.
I THINK WE CAN SURMISE THAT THE ORGANIC MATTER FROM THE VAULT HAS IN THE VERY LEAST SERVED AS A CATALYST... AN EVOLUTIONARY ACCELERATOR...
RRRRRRRUUMMBLE
HEY!
QUAKE?
11

It was one of those vast herds of sambuks that thunder across the interior. They broke and flowed around the fires, a surging river of muscle and bone...
As was bound to happen, an ancient, crippled bull, unable to match the herd's momentum, was forced through the fires...
KEEP HIM FROM THE POOL!
12

Unfortunately, Jack's best efforts proved futile.
At the edge of the pool, the crazed sambuk's legs crumbled ...
I suppose we were lucky the rest of the herd passed without incident.
But the damage had been done. External factors in the shape of a 30 ton rotting sambuk carcass had *drastically* changed the *environment* of our pool...
13

As we had predicted, the unique life force in the waters responded immediately to the sudden flood of nutrients released from the decomposing beast.
Within two days, a completely new and sophisticated scavenging ecosystem had replaced the old. The pool now teemed with highly evolved life... some of it air-breathing.
WE'VE GOT TO FIGURE THAT THE SAMBUK CARRIED IN EGGS... SEEDS...
SOME OF THE INSECTS, MAYBE... THE BLOWFLIES, YES... BUT NOT ALL THIS AQUATIC LIFE... THERE'S TOO MUCH DIVERSITY... IT'S DEVELOPED MUCH TOO QUICKLY...
NO... WE MUST CONCLUDE THERE IS AN EXTREMELY ABNORMAL FORCE AT WORK HERE...
LET'S LIST THE KNOWN FACTS:
ONE... WE UNLEASH AN ORGANIC, PLASMIC ENTITY THAT HAD BEEN PRESERVED FOR CENTURIES...
TWO... SAID ENTITY ACTIVELY SEEKS OUT WATER... THE ENVIRONMENT MOST CONDUCIVE TO THE DEVELOPMENT OF PRIMITIVE LIFE...
14

Who can explain what happened next?
Maybe the fuel disturbed some nest in the choking vegetation...
Hannah doused the torch and the attack on Tenrec ceased...
Although I abhor violence, I had decided this madness had gone on long enough...
I stepped out from where I'd been hiding and...
OOH... DID YOU HAVE TO HIT HIM SO HARD?
19

Tenrec had suffered a concussion, and some nasty *bites*.
For his own safety, and despite his noble *protestations*, we tied him down and sent him home.
Free from Tenrec's interference, we could begin our studies in earnest.
Since we became *organized*, samples of our unique waters have been distributed to a number of scientific stations.
The situation is now *well* out of Tenrec's control.
Just last week our marine laboratory reported observing the development of a primitive *notochord!* ...Can vertebrates be long off?
I think that Tenrec will be OK, however...
Instead of remaining with us at the pool, searching for answers to the questions that obsess her, Hannah chose to accompany him home.
What she sees in him escapes me.
THE END
20

CROSSED CURRENTS
WHILE HIS CREW FINISHED TYING UP HIS DECREPIT BOAT, GNEISS WATCHED THE OTHER RETURNING FISHERMEN WITH HIS USUAL MIXTURE OF ENVY AND CONTEMPT.
HE ENVIED THEIR SIZEABLE CATCHES, THE WEIGHT OF WHICH CAUSED THEIR SLEEK VESSELS TO RIDE SO LOW IN THE WATER, BUT HE FELT NOTHING BUT CONTEMPT FOR THE LONG HOURS OF BACKBREAKING LABOR NEEDED TO PRODUCE SUCH A HAUL.
Steve Stiles
THE HARBORMASTER APPROACHED GNEISS AND HIS CREW AS THEY WERE LEAVING...
THE SAME CATCH AS ALWAYS, EH, GNEISS? AND YOUR BOAT STILL UNREPAIRED...
YOU'RE A LAZY MAN, GNEISS, AND IF YOUR CATCHES DON'T IMPROVE SOON, YOU'RE LOSIN' YOUR SHIP'S BERTH... THERE ARE LOTS OF OTHERS WAITIN' FOR YOUR SLIP AND YOUR JOB...
1
DON'T COUNT ME OUT YET, YOU SELF-RIGHTEOUS OLD CHUM BUCKET!
MAYBE YOU'LL BE SEEIN' SOME BIG CATCHES FROM ME SOONER THAN YOU THINK... MAYBE THE BIGGEST HAUL YOU'VE EVER SEEN!
BUT IT'LL BE DONE MY WAY! I'LL BE USIN' MY BRAINS INSTEAD OF SWEATIN' MY BUTT OFF!

GNEISS... WHAT YA SAID TA DAT OLD WINDBAG... ARE YA HIDIN' SUMTHIN' FROM US? YA GOT SUMTHIN' UP YER SLEEVE?
SHUT YER TRAP, MAKO! I WAS GONNA SHOW YOUSE ANYWAY...
I'M GONNA SHOW YOUSE SUMTHIN' I AINT SHOWN NOBODY BEFORE...
WELCOME TO MY LABORATORY. BETCHA DIDN'T KNOW I HAD ME A SCIENTIFIC CAREER BEFORE... AH... CIRCUMSTANCES FORCED ME TO TAKE UP FISHIN'...
BACK IN THE DAYS I WORKED DOWN IN THE CATACOMBS, I STUDIED UNDER SCHARNHORST HERSELF!
WHY, I EVEN MET THE GREAT DR. FESSENDEN!
2
AND I MADE THE RIGHT CONTACTS TO INSURE I COULD GET SOME OF THIS HERE PRE-CATACLYSMIC EQUIPMENT FOR MY OWN USE...

NOW PAY ATTENTION, THIS IS IMPORTANT...
THIS HERE DEVICE IS CALLED A BIO-VOLTAMMETER. MEASURES ELECTRICAL IMPULSES DISCHARGED BY MARINE CREATURES...
ALL CREATURES GIVE OFF MINUTE ELECTRICAL IMPULSES... THIS HERE FISH IS HEALTHY AND, AS YOU CAN SEE, GIVES OFF A STEADY, REGULAR PULSE...
NOW I'M GONNA GIVE YOUSE A BIOLOGICAL LESSON...
FWAP!
WHAT I HAVE CREATED HERE IS A DISTRESSED FISH...
...A FISH WHICH EMITS SOME VERY SPASTIC IMPULSES...
...THESE ERATIC IMPULSES ARE JUST WHAT GET A PREDATOR'S ATTENTION... THEY CAN SENSE AN EASY MEAL...
3

...AND IT SEEMS MY LI'L SEA VIPER HAS TAKEN NOTE OF THE SITUATION...
NOW WATCH THIS! THIS IS THE PART THAT REALLY CONCERNS US!
THE SEA VIPER'S WEAPON IS ELECTRICITY!
LOOK AT THE VOLTAMMETER! THE VIPER JUST KILLED THE FISH WITH A MASSIVE DOSE OF ELECTRICITY!
CAN YOUSE SEE HOW THE VIPER'S HUNTIN' TECHNIQUE COULD BE A BENEFIT TO OUR FISHIN' CAREERS?
UH...
NEVER MIND, YA SPONGEBRAINS!
I'LL TAKE CARE OF THE THINKIN'...
AS A MATTER OF FACT, I'VE ALREADY DONE US A LOAD OF THINKIN', AN' HERE'S THE RESULT...
IT'S TAKEN ME A LONG TIME TO WORK OUT THE BUGS, BUT NOW IT'S TIME TO GIVE MY SUPER-GENERATOR A WORKOUT...
...TOMORROW WE DO US SOME LABOR SAVIN' HIGH VOLTAGE FISHIN'...
4

THE NEXT MORNING DAWNED TO FIND GNEISS' BOAT LAGGING WELL BEHIND THE REST OF THE FLEET, AS USUAL...
I DUNNO... THERE'S SUMTHIN' ABOUT DIS HERE DEAL DAT KINDA BOTHERS ME...
I TOLD YOU TO SHUT UP AN' LET ME DO THE THINKIN'!
GNEISS TOOK THEM OUT FAR PAST THE FLEET, INTO DEEPER WATERS THAN WERE NORMALLY FISHED...
I DON'T WANT NONE OF THEM OTHERS FINDIN' OUT ABOUT MY WUNNERFUL NEW SCIENTIFIC TECHNIQUE...
CAREFUL WITH THEM LINES...
ALL THE LINES PLAYED OUT?
NOW STAND BACK FROM 'EM, 'CAUSE MY SUPER-GENERATOR CONVERTS A WHOLE MESS O' ELECTRICITY FROM SEA WATER AN' I'M TURNIN ON THE JUICE!
WITHIN MINUTES GNEISS' INVENTION HAD BECOME A PROVEN SUCCESS...
I KNEW IT! I KNEW IT'D WORK LIKE A CHARM! IT ZAPPED 'EM REAL GOOD!
5

GENERATOR'S OFF!
YA CAN START COLLECTIN' OUR FIRST HARVEST, BOYS...
THROW BACK THE ONES WE CAN'T USE...
DIS AINT TOO BAD...
PULL IN THE LINES AND WE'LL MOVE ON... WE'RE GONNA SHOW THE OTHERS A CATCH LIKE THEY'VE NEVER SEEN!
6

OVER AND OVER, TIME AFTER TIME, THE DEADLY CURRENT SURGED, CLOGGING THE SEA WITH DEAD AND DYING MARINE LIFE...
UNTIL...
CHOKE!
THEY'RE EATIN' OUR FISH!
...GNEISS... LET'S GET OUTTA HERE!
THEY'RE EATIN' OUR FISH! I'M GONNA FRY THEM TOO!
7

GNEISS THREW THE SWITCH, THE ELECTRICITY SURGED AND... THE SEA VIPERS SLOWLY BEGAN TO SWIM TOWARD THE BOAT!
FOR GOD'S SAKE, GNEISS... TURN IT OFF!
WE'RE ATTRACTIN' EM... JEST LIKE DAT FISH IN YER TANK!
JOHN B.
BUT IT WAS TOO LATE...
THEY FOUND THE DERELICT SEVERAL DAYS LATER. THERE WAS NOTHING ON BOARD SAVE SOME USELESS, BURNED OUT ELECTRICAL APPARATUS.
NO ONE WAS PARTICULARLY SHOCKED TO LEARN WHO THE VESSEL HAD BELONGED TO...
END
8

IN THE DREAMTIME...
Deep, deep down...at the dark, soft bottom of a lake somewhere in the interior... something stirs in the muck...
Responding to some distant and imperceptible pull, tiny bubbles work free of the depths and begin to rise...
Slowly, slowly they rise and expand until they break the surface and rupture, combining their poisonous contents into a deadly, rolling cloud.
...SO, IF THE ...UH... CARBON ACCELERATOR DEFRIBULATES, I WANT TO...AH...FIRST CHECK THE TORQUE ON THE...UH...GERSCHLÄNG SPROKKET...
MARK SCHULTZ © 1989
1

NO, NO, NO! PAY ATTENTION!
IF YOU WANNA LEARN TO DRIVE, FIRST YA LEARN TO BE A MECHANIC. YOU'RE NOT ALWAYS GONNA HAVE ME AROUND TO FIX YOUR WAGON...
WELL, ISN'T THIS COZY?
YOU'VE RECOVERED FROM YOUR CONCUSSION VERY WELL, HAVEN'T YOU, JACK?
THANKS, NO DOUBT, TO YOUR BEDSIDE MINISTRATIONS, AMBASSADOR DUNDEE.
OH, FOR CRYIN' OUT LOUD, DAHLGREN!
WHAT ARE YOU DOIN' HERE? I NEED PEACE AND QUIET IN MY FRAIL CONDITION...
STUFF IT, TENREC! OR I'LL KICK YOU IN THE TAIL AND KNOCK OUT YOUR SECOND BRAIN.
MY, YOU'VE DONE WONDERS FIXING UP THIS LI'L OL' BEDROOM...
UH... MAYBE I'D BETTER GO...
WHY, YES, AMBASSADOR DUNDEE...
YOU'RE SUPPOSED TO BE TEACHING OUR FISHERMEN SOMETHING, AREN'T YOU... HMMMM?
YOU ALWAYS WERE A CHARMER...
WHAT THE HELL'S ALL THIS ABOUT?
WHILE WE'RE ALONE, LISTEN UP.
BULGAR PRESENTED A REPORT TO THE COUNCIL YESTERDAY ABOUT YOUR INTERFERENCE AT THAT BURIED VAULT...AND WITH WHATEVER IT IS HE FOUND IN IT.
AS YOU CAN IMAGINE, THE OTHER GOVERNORS TAKE A DIM VIEW OF YOUR ACTIONS...
2

AM I SUPPOSED TO BE SURPRISED?
IF YOU PEOPLE HAD ANY IDEA WHAT BULGAR IS PLAYIN' AROUND WITH...
SHUT UP AND LISTEN, YA BIG LUG.
WE BOTH KNOW THAT THE COUNCIL HAS NEVER COME DOWN TOO HARD ON ITS LOOSE CANNON...
...BUT THIS TIME MAYBE YOU'D BETTER LAY LOW FOR A WHILE...
...CERTAIN POWERFUL... ER...ELEMENTS HAVE BECOME EXTREMELY HOSTILE TOWARD YOUR ATTITUDES...
BULGAR?
NO... HE'S GOT NOTHING TO DO WITH THIS. I CAN'T SAY ANYTHING MORE NOW... TRUST ME!
YOU KNOW THE RISKS I RUN TO KEEP YOU INFORMED. IF THE COUNCIL EVER FOUND OUT...
I KNOW... HOW CAN I EVER REPAY YOU?
HA HA! MAYBE COOL YOUR DIPLOMATIC RELATIONS WITH THE WASSOON...
AHEM...
MUSTAPHA CAIRO! WHAT GIVES? DOESN'T ANYONE KNOCK ANYMORE?
HEY, THE DOOR WAS OPEN.
I'VE GOT A FAVOR TO ASK, JACK...
THIS'LL DO YOU SOME GOOD...
I'VE GOT A ROAD GANG OUT IN THE LAKE REGION SMOOTHING OUT SOME VOLCANIC STUFF.
THAT FIGURES.
THEY'VE FINISHED THE STRETCH I LEFT 'EM WITH, AND ARE WAITING FOR ME TO GET BACK...
BUT I'VE BEEN TIED UP WITH THE BRIDGE PROJECT AT GRISSOM GULCH...
3

Deep, deep down...at the dark soft bottom of a lake somewhere in the interior...something... stirs in the muck...
I'D BE MUCH OBLIGED IF YOU'D GET THESE DIRECTIVES TO THE CREW TODAY... BEFORE THEY GET RESTLESS.
AND TELL THEM I'LL BE ALONG IN A DAY OR TWO...
WELL, IT SEEMS LIKE NO ONE CAN STAND TO SEE ME TAKING IT EASY. DAHLGREN THINKS I CAN USE A LITTLE FRESH AIR, TOO...
HANNAH! GET OVER HERE! YOU READY FOR YOUR FIRST LESSON BEHIND THE WHEEL?!
OK... YOU GOT YOURSELF A RUNNER, MUSTAPHA.
THANKS, MUSTAPHA. THAT WAS EASIER THAN I THOUGHT IT WOULD BE.
4

... NOW TRY PUTTING HER IN FOURTH...
...CLUTCH IN... SHIFT... CLUTCH OUT SLOWLY WHILE YOU GENTLY GIVE HER THE GAS...
WHUMPF!
LIKE THAT?
SORT OF LIKE THAT.
WHAT'S THIS? WHAT ARE "TURN SIGNALS" FOR?
I'VE GOT NO IDEA.
Something stirs and begins to work free...
5

WHAT IS MUSTAPHA BUILDING IN THIS LAKE REGION?
A ROAD... A SHORTCUT TO THE COPPER MINES. IF HE CAN CARVE THIS ROUTE, WE'LL CUT TWO DAYS OFF THE CALHOON RUN.
LET'S SEE THE MAP...
THIS IS GEOLOGICALLY ACTIVE EARTH... IT GETS MOUNTAINOUS...
WE SHOULD BE SEEING SOME ROADWORK SOON...
MUSTAPHA SAYS WE'LL FIND THE CREW CAMPED BY MACHTANDO. THAT'S THE SECOND LAKE WE'LL SEE.
THĒU
MACHTANDO
ALWON
ORENDA
YEP, THIS IS A ROAD.
WE HUMANS JUST CAN'T SEEM TO ACCEPT THINGS AS THEY ARE, CAN WE? IF IT'S INCONVENIENT, WE CAN ALWAYS REMAKE IT TO OUR LIKING. HA!
LOOK AT THIS! THEY'VE BLASTED RIGHT THROUGH THESE ROCKS!
SO THIS IS A ROAD!
I TELL YOU, HANNAH, THESE ARE STRANGE TIMES...
IT SCARES ME... I SEE MY TRIBE STARTING TO MAKE THE SAME MISTAKES THE ANCIENTS MADE...
LAND SHAPING ...AND FARMS! GOOD LORD! WE NEED FEWER PEOPLE, NOT MORE AGRI-CULTURE!
WHY CAN'T THEY SEE WHAT THEY'RE DOING?
I CAN'T SEEM TO INFLUENCE 'EM ANYMORE...
SO LET BULGAR CONTINUE WITH HIS EXPERIMENTS... I CAN'T FIGHT A WAR ON ALL FRONTS.
HEY, THIS IS ORENDA... THE FIRST LAKE.
6

THERE'S MACH-TANDO.
WE OUGHT TO SEE THE CAMP SOON... I WONDER WHY THEY DON'T HAVE LOOKOUTS POSTED?
I THINK I SAW SOMETHING BACK...
CHOKE
OH MY...
STOP THE CAR.
7

WHAT?..
I DON'T KNOW.
NO SIGNS OF STRUGGLE... NO ATTACK...
THEY'VE BEEN DEAD A DAY AT LEAST...
LOOK... FISH, BIRDS, SLITHERS ...
THERE WAS A MASS KILLING HERE.
8

NEVER SAW ANYTHING LIKE THIS.
IT'S GETTING DARK... THERE'S NOTHING WE CAN DO NOW.
LET'S SET UP CAMP. BUT NOT HERE.
"Let's head back to Orenda."
Orenda...
WHAT CAN YOU TELL ME ABOUT THESE LAKES?
WELL... THEY'RE VOLCANIC CRATERS THAT TAP GROUND WATER. THEY VARY IN DEPTH...
HOLD ON A MINUTE...
... I WANT TO POINT THIS BUGGY IN THE RIGHT DIRECTION.
YOU NEVER KNOW... WE MIGHT HAVE TO LEAVE IN A HURRY.
9

THE MOONS ARE FUSED.
THEY'RE IN ALIGNMENT ALL THIS MONTH.
DID YOU KNOW THAT BEFORE THE CATACLYSM THE EARTH HAD ONLY **ONE** SATELLITE?
GOOD FOR THE ANCIENTS.
TWO MOONS KISSING... ALWAYS BAD MEDICINE... ALWAYS SPELLS DISASTER...
DON'T BE RIDICULOUS.
TELL ME MORE ABOUT THIS REGION.
THE LAKE DEPTHS VARY GREATLY. MACHTANDO'S VERY SHALLOW, AND ORENDA HERE IS RELATIVELY DEEP.
LOTS OF HOT SPRINGS AND FUMAROLES IN THE AREA...
NONE OF THAT CAN EXPLAIN WHAT HAPPENED...
YOUR PROBLEM IS YOU NEVER LOOK **PAST** YOUR SCIENCE.
I THINK THE ROAD CREW'S BLASTIN' AND GOUGIN' PROBABLY OFFENDED THE **SPIRITS** OF THE LAKE...
CRUNCK
WHAT WAS THAT?!
SOMETHIN' BIG!
DON'T SHOOT... IT'S JUST A HORNBILL ON IT'S WAY TO THE SHALLOWS FOR THE NIGHT.
10

SOME WOODSMAN YOU ARE! YOU CAMPED US RIGHT ALONGSIDE A HORNBILL PATH.
SO THEY'LL KEEP US FROM OVERSLEEPING...
TAKE THE FIRST WATCH, WILL YOU? I'VE GOT A LOUSY HEADACHE...
SURE, SURE... GET SOME SLEEP. TOMORROW'S GONNA BE BRUTAL.
NOT FEELIN' SO GOOD MYSELF...
ZZZZZZZ...
11

I'm dreaming. I know that I'm dreaming. That usually means the grith have something to tell me.
But I can't sense them anywhere.
I'm under water, but I can breathe. I'm a...fish.
In the distance I see watery figures approaching...
Suddenly, they loom over me. They're gigantic, soft, formless. They've come for me.
They are the ancestors of all men, of all grith, of all fish, and they want me.
I'm terrified... I can't possibly do what they expect of me.
I turn and flee and...
I'm caught in a net. I struggle to no avail...
12

I walk along a seashore with Jack, hand in hand. Something wonderful is about to happen...
I'm dreaming. I know I'm dreaming.
Jack leads me into the surf... into the waves.
I knew this was coming.
We float under the waves. I look at Jack, and...
THERE IS MUCH YOU NEED TO KNOW, HANNAH DUNDEE, AND LITTLE TIME.
LIE DOWN... **HERE.**
SOON YOU WILL BE GOING ON A GREAT JOURNEY... DOWN, DOWN, DOWN... AND WE WILL PREPARE YOU FOR THAT.
I do as he says. I'm ready.
BUT NOW YOU MUST PAY ATTENTION AND UNDERSTAND THE **IMMEDIATE** DANGER...
13

I'm yanked from the water and hang in the air and twist...
...as Hannah inspects me.
She looks back and there they are...the ancestors.
THROW HIM BACK. THERE IS SOMETHING HE MUST LEARN.
No! Not back into the water...
Not into the cauldron!
14

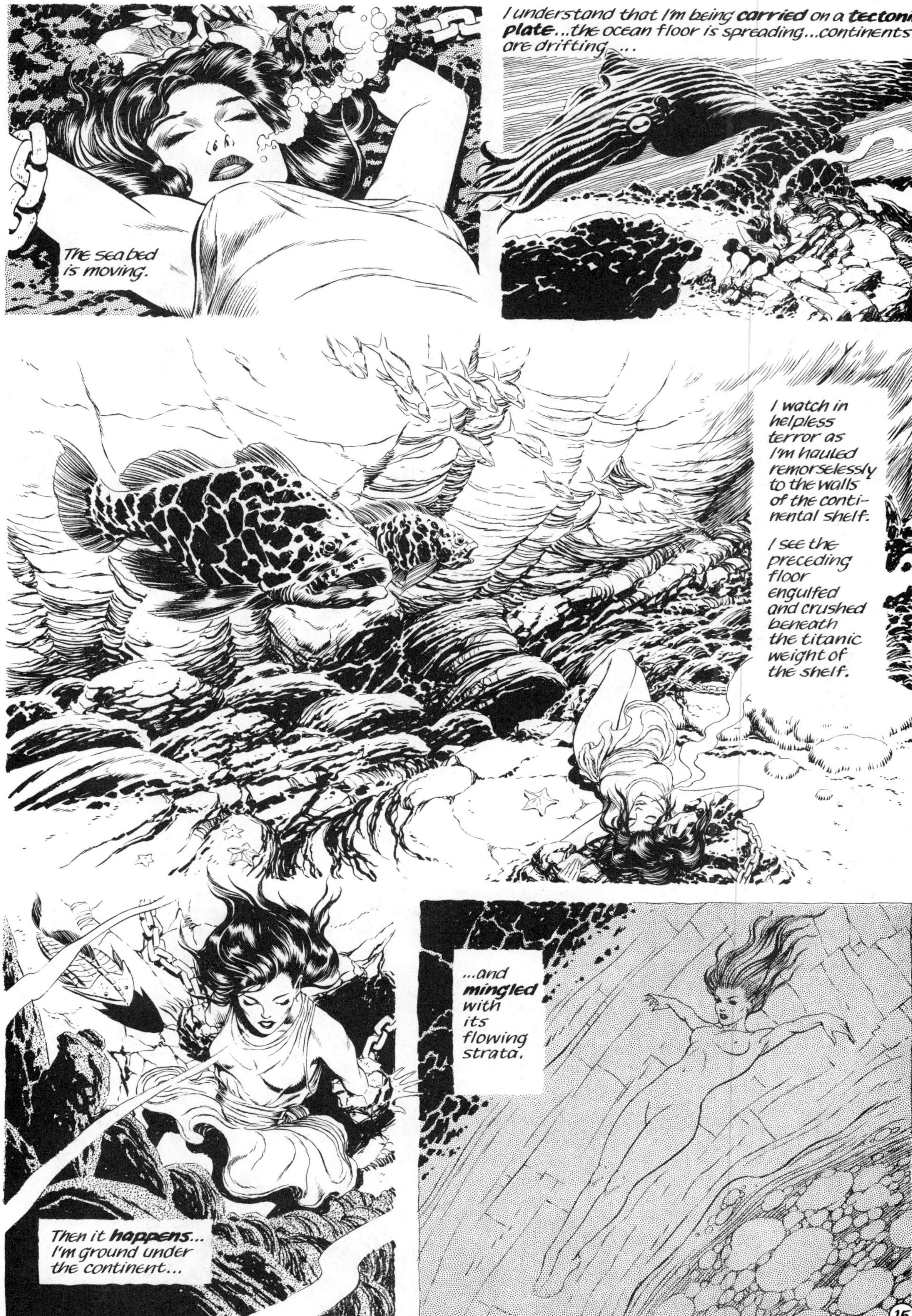
The sea bed is moving.
I understand that I'm being **carried** on a **tectonic plate**...the ocean floor is spreading...continents are drifting...
I watch in helpless terror as I'm hauled remorselessly to the walls of the conti-nental shelf.
I see the preceding floor engulfed and crushed beneath the titanic weight of the shelf.
Then it **happens**... I'm ground under the continent...
...and **mingled** with its flowing strata.
15

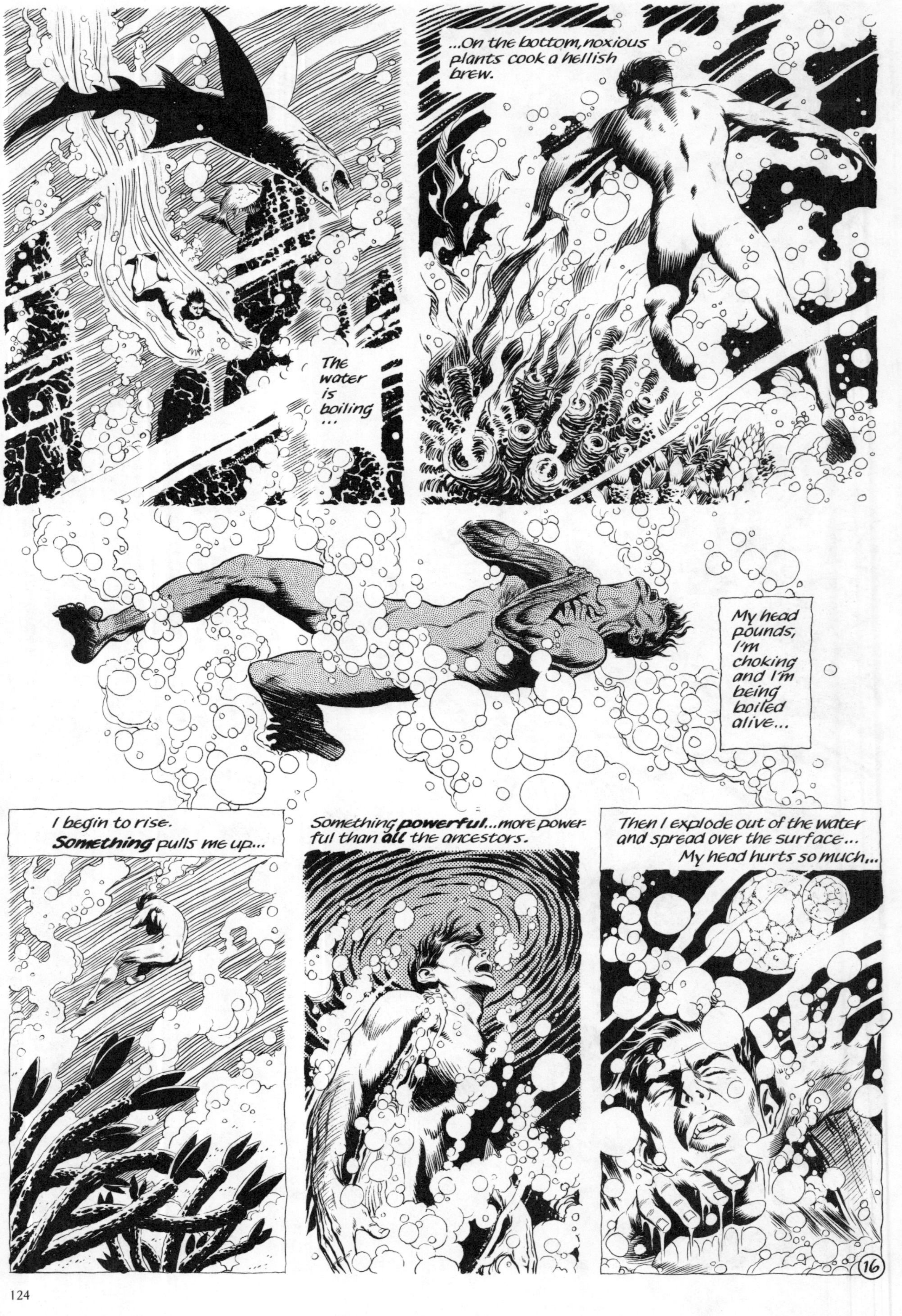

The water is boiling ...
...On the bottom, noxious plants cook a hellish brew.
My head pounds, I'm choking and I'm being boiled alive...
I begin to rise. **Something** pulls me up...
Something **powerful**...more powerful than **all** the ancestors.
Then I explode out of the water and spread over the surface...
My head hurts so much...
16

The strata and I climb up and up into the land...
...until we meet and fuse with a rising river of magma.
We hit ground water and I'm released with a stream of **poisons** into the boiling liquid.
The poisons and I leach upward...into the bottom of a freshwater...**lake**.
I know where I am.
The poisons and I rise towards the surface...
My head pounds ... The water boils ...
We burst and spill over the surface ... I feel the moons tearing at my brain...
17

JACK! WAKE UP! WE GOTTA...
CHOKE
... GET OUT OF HERE... COUGH, COUGH
...CAN'T BREATHE... COUGH
...GET TO THE CAR...
HRRONK!
GAK!
18

BRMM BRMM
VVRRROOOOOM
SKRRR
REEEEEE
KER-RUMF!
19

"We found them on our way out to meet the road gang..."
"They were in bad shape... retching and heaving... fighting for every breath..."
"After a few hours they recovered enough to recount what had happened.
"We listened in horror as they told of the fate of the road crew, and we marvelled at the luck that had delivered them from a similar end."
"Hannah offered a theory that the combined gravitational pull of the aligned moons disturbed a layer of poisonous gases which had accumulated at the bottoms of those volcanic lakes..."
"Jack just smiled that smug smile of his.
"He's a strange one. He may regret the deaths of the road gang and all, but I'm sure he was relieved that the earth had reacted against a perceived intrusion.
"I think he felt reassured."
THE END
20